Toussaint,
The Greatest

Episode 3

MUFFIN
WARS

#Team JT

#Team Muffin

Plum Street Press

A Division of Yes, MAM Creations

Contents

Prologue _____ vii

Chapter 1: *I ♡ NY* _____ 1

Chapter 2: *Muffin Mania* 10

Chapter 3: *Plan B* ~~~~~~~~~~~~~~~~~ 20

Chapter 4: *Game On* 29

Chapter 5: *The Absolute Worst* ⅢⅢⅢⅢⅢⅢⅢⅢ 39

Chapter 6: *The Apprentice* ~~~~~~~~~~~~ 49

Chapter 7: *The Dynamic Duo* ═══════ 55

Epilogue ════════════════════════ 59

Who Made This Book? ⬦⬦⬦⬦⬦⬦⬦⬦⬦⬦ 60

To older girl
and younger
boy.
May you
always
be the
greatest.

Prologue

JADEN TOUSSAINT

Specializes in: Knowing Stuff. And also, ninja dancing. He's really, really good at ninja dancing.

OWEN

Jokester and action expert.

THE FRIENDS

Extreme dinosaur safari
bungee jumping?
Owen is your guy.

EVIE

Don't let the cuteness fool you.

This girl packs a punch.
Excels at: Being in
Charge.

SONJA

Also draws
excellent rainbows.

Cicada hunter and math whiz.

WINSTON

Can quote stats
from every World
Cup final.*

*that he has been alive for

Spots hurt feelings and distracted goalies from miles away.

Past president, International
Society of Stealthy Felines.
Resigned in scandal. Likes
to be held like a baby and
scratch stuff.

GRANDMASTER, CAT CHESS.

GRIS
GRIS

ANIMAL OF MYSTERY

Guinea Pig never has the same name
two weeks in a row.

?????

This week you can call him One!
(Singular Sensation)

Baba:

Tall. Competitive. Competitive about being tall. Gives great piggyback rides. Prefers to be called "baba," which means "father" in Swahili. Does not speak Swahili.

Mama:

Loving. No nonsense. Most often seen reading fantasy books or experimenting with bean desserts. Gives good hugs.

Sissy:

Reader. Writer. Animal lover. Once gave up meat for 6 months, but was broken by the smell of turkey bacon. Plans to be the first PhD chemist to star in a Broadway musical.

Chapter 1
I ♥ NY

Jaden Toussaint loved New York.

Sometimes [gasp] JT loved New York even more than his hometown. That was really saying something, because JT's hometown was really special. But how could he help it when New York had so much to love?

New York had giant statues with spiky crowns (well, the one giant statue with a spiky crown, but it was really big), pretzels as big as your head, and cheese pizza on pretty much every corner.

3

Jaden Toussaint had been to lots of other places...the beach...his uncle's farm in Texas...and the biggest, fanciest house

you ever saw in North Carolina, but none of that could even hold a candle to New York.

There were only two bad
things about New York:

1. JT's mama had
insisted on correcting
him every time he asked
to climb the mountains
in Central Park.

2. There was an awful lot of
stuff UNDERground. There
were even trains that ran
underground.

There were only worms and
water underground where

Jaden lived, so basements and subways and
all the other underground things were a little
creepy to him. And cool. Ok, creepy-cool.

He added Underground Stuff to his list of Cool
New York Facts. His list had gotten pretty long
ever since he found out his cousin Muffin was
coming to visit.

6

Muffin's real name was Makaela. She and her parents were usually so busy traveling all over the world that Muffin and JT had only met once, and JT could barely remember it.

  Pause.

Technically, he couldn't remember it at all, but he had seen lots of pictures and that was pretty much the same thing, right?

  Unpause.

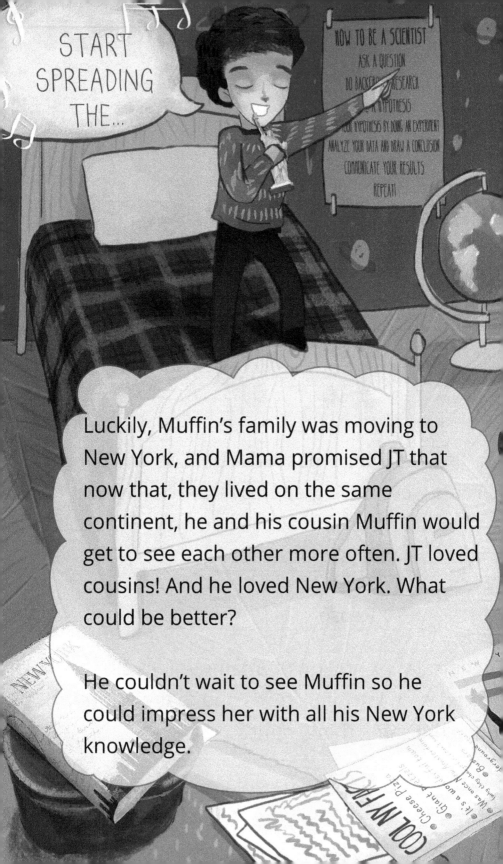

Luckily, Muffin's family was moving to New York, and Mama promised JT that now that, they lived on the same continent, he and his cousin Muffin would get to see each other more often. JT loved cousins! And he loved New York. What could be better?

He couldn't wait to see Muffin so he could impress her with all his New York knowledge.

DING
DONG

He had strategically placed a few mementos and pieces of memorabilia around the house to make sure New York came up in conversation. You know. Casually. Muffin would be staying at his house for three whole days while her parents went apartment hunting in New York, so that gave them plenty of time.

JT was so excited that he almost burst into song! But before he did, the doorbell rang.

Chapter 2
MUFFIN MANIA

Blueberry

Mandarin Orange

Apple

Pear

Strawberry

Blackberry

Sneaky Banana

Jaden Toussaint scrambled to the door so quickly that he almost forgot about the muffins! Even though he knew that Muffin and her parents wouldn't arrive until evening, and even though he knew muffins were a breakfast food,

Jaden Toussaint had spent all afternoon baking muffins all by himself. Well, mostly all by himself. Sissy had helped a little with the oven, and the blender, and the mixer, and getting the pan from the high shelf. And maybe she had cracked some of the eggs and made sure that the ingredients were measured just right, but that was it. He had done all the rest by himself and they were his most delicious muffin invention yet: Super-Sneaky Seven Fruit Fruit Salad Muffins.

If one kind of fruit was good, seven kinds of fruit would be even better, right? And when you're making muffins for a person NAMED Muffin, you really have to go all out.

By the time JT got to the living room with his basket of muffin-y goodness, everyone seemed to already be hanging out without him. And what was worse was they seemed to be having a good time. Without him.

Before JT could panic, Mama held out her arms and he ran to sit on her lap. Mama and Muffin had been talking, and as he settled onto Mama's lap Muffin turned to him and said, "Salut, cousin." But to Jaden it sounded like she said, "Salad Cruising," which is a pretty weird thing to say to someone when you first meet them. But before he even had a chance to realize that she was speaking French to him, Muffin said, "Oh. I'm sorry. I forgot you only speak English. I said, 'Hi, cousin.'"

"I know," Jaden Toussaint said.

       Pause.

He didn't really know. At least, he didn't know it right away. But he would have figured it out if he had just waited a second. JT didn't know a whole lot of French, but he did know how to say hello if she had just given him a chance to think. It all happened so fast that "I know" was the first thing that popped out. He wished he had said, "Bon soir."

  Unpause.

Before the night was over, Muffin had
impressed everyone in the house. She had
spoken French with Mama, geeked out with
Sissy about some engineer set, and wowed
Baba with her knowledge of obscure sports.
Everyone was going crazy for Muffin, and not

a single person had tasted JT's muffins, not even at dessert.

To top it all off, Muffin had monopolized the conversation so much that he hadn't had a chance to impress anyone with his knowledge of animal facts or New York.

By the time he realized that he hadn't had a chance to be his usual, informative self, his aunt and uncle were already on their way back to the airport and he and Muffin were headed to bed.

Not to worry. Once they got to his room, Muffin couldn't help but notice the few NY items he had strategically placed around the room. Once Muffin saw all his cool NY stuff, JT thought she would be so impressed that they'd stay up all night talking about New York and becoming lifelong friends.

Unfortunately, it didn't turn out that way.

After he had climbed into bed, Mama came in to tell him that Muffin was too sad about her parents being gone to sleep in a sleeping bag on the floor in JT's room.

Ah, well. His plan to impress his cousin so she would think he was awesome and become his best friend in the whole wide world hadn't worked. It was a bit disappointing but, as a scientist, Jaden Toussaint understood that your first plan doesn't work every time. That's why he always, always had a back-up plan.

Chapter 3
PLAN B

Jaden Toussaint woke the next morning with pep in his step. After all, he had 3 things going for him:

1. He was awesome.

2. Muffin was going to go to school with him, a place where he was often his most awesome.

3. School contained one of his most powerful secret weapons: Chinese.

The night before, while everyone else was sleeping, JT had stayed up late doing research. First he made a list of all the places Muffin had lived. Then he used the atlas to check which language each country spoke. Luckily, none of the countries spoke Chinese, and it just so happened that every Tuesday Jaden Toussaint's class started with a Chinese lesson from Mr. Zhou. It was perfect! Muffin had caught him by surprise with French, but now he would catch Muffin by surprise with Chinese.

PLACES MUFFIN HAS LIVED	LANGUAGES
Monrovia, Liberia	Bandi, Bassa, Dan, Dewoin, English, Gbii
Cairo, Egypt	Egyptian Arabic
Rio, Brazil	Portuguese
Dubai, United Arab Emirates	English, Arabic
Paris, France	French

Problem: Plan B was a flop from the very beginning.

First, Ms. Bates had made a big deal about introducing Muffin to the class.

That part he expected. He had even written it down in his plan. What he didn't bank on was that when Muffin stood up to say hello, she didn't just stop at hello. After hello, she added:

He had suspected she might say hello in French instead of English, French being one of her superpowers. But saying hello once in English and twice in French was a little bit of a curve ball. Still, nothing JT couldn't handle.

Then another one slipped out.

And another one.

And another one.

And before he knew it Muffin was spitting out a string of hellos so long and smooth that the whole class erupted into applause.

JT clapped, too. He was pretty sure that one of those hellos was in Martian. Or Mermish. Whatever it was, it was impressive. When she went to sit down, several kids tried to make room for her to sit next to them.

TALA.

25

He had not counted on Muffin being quite so spectacular. But scientists expect set-backs, and Jaden Toussaint was nothing if not a good scientist. None of those hellos had been in Chinese. Maybe, he thought hopefully, maybe since she obviously likes languages she'll be even more impressed when she realizes I can teach her Chinese.

    Pause.

Jaden Toussaint couldn't really teach Muffin Chinese. Well, at least not ALL of Chinese. But he knew the names of fruits and his colors and several songs including one about two tigers. He even knew all the numbers through 10. That counted, right?

    Unpause.

Mr. Zhou joined the circle with his guitar.

Jaden Toussaint's hand shot straight up, but before Mr. Zhou could call on him, Muffin just blurted out, "Do you mean Liang Zhi Lao Hu?"

What? No!! Not Chinese!!!

Jaden Toussaint was so upset to hear
Chinese coming out of Muffin's mouth
that he forgot to be upset about the fact
that she had called out an answer without
even raising her hand. Chinese was *his*
thing. *He* was good at Chinese. So
while everybody else sang Liang Zhi Lao
Hu, Jaden Toussaint wracked his brain for
a back-up plan to his back-up plan. One
thing for sure: he was going to need back-
up.

Chapter 4
GAME ON

At recess, Sonja took Muffin off to hunt for cicada shells. He could have been bothered by the fact that it had only taken about three seconds for Muffin to make friends with one of his friends, but he decided against it. They could both be friends with Sonja. And besides, it gave him a chance to talk to Winston, Owen, and Evie about his problem without his problem standing right there.

"She does seem really cool," said Owen.
"That's not helping, Owen," said Evie,
giving Jaden Toussaint a sympathetic
look. "I mean, she does seem really cool.
Really, really cool. But she's not cooler
than JT."

"I don't know," said Owen. "Did you hear her doing math facts? I think she got all the way to multiplication."

"That's not helping, Owen!" Evie repeated.

"Wait. Did you seriously hear her saying her multiplication facts? That's amazing!"

"Guys," Winston interrupted, "we're coming at this from the wrong angle. I think if we just stop and take a moment to meditate we can--"

"Winston, we don't have time to meditate. Recess is almost over and Muffin is getting cooler by the minute. We have to take action," Evie declared.

Evie was right. A lot of other cicada hunters had joined Muffin and Sonja, and it looked like Muffin was giving them some kind of presentation?

Wait a minute. Was she teaching them animal facts? Not animal facts, too?!? Was nothing sacred? Something had to be done.

"What should I do, Evie?" JT asked.

The teachers had started to move to the line up area. Whatever they were going to do, it had to be fast.

"You just have to beat her at something."

"But what? She's good at everything."

"Kindergarten chess," said Evie.

Jaden Toussaint smiled.

Kindergarten chess was like regular chess, except trickier because all the pawns were actual kindergartners who might wander off and decide to play something else in the middle of your game. But it just so happened that Jaden Toussaint was the reigning world champion of kindergarten chess. Even if Muffin was good at ordinary chess, there was no way she could beat him at kindergarten chess.

"Evie, you're a genius!" JT exclaimed just as the teachers blew the first whistle.

Now, everybody at JT's school knew about the two whistle system. The first whistle meant that everybody on the playground had to freeze. The second whistle meant that it was time to walk, not run, walk to your line up spot.

The second whistle blew. JT started to walk, he really did, but when he caught sight of Muffin running toward him, he was so caught up in the idea of beating her at something—anything—that he completely forgot about the two whistle system.

Truthfully, in that moment, he forgot about pretty much everything. The only thing he could think of was that Muffin seemed to be racing him and he needed to win.

As if by magic, every lesson Baba had ever tried to teach him about running instantly clicked into place.

Head upright. Eyes straight ahead. Chest lifted. Slight lean forward. Arms at 90 degrees. Hands loose.

His heels barely touched the ground. Muffin was gaining on him, but it was too late. As he picked up speed he got even more relaxed and he knew he must be moving as fast as a cheetah. If not a cheetah, at least as fast as Usain Bolt.

Almost there.

Almost...

He made it! He made it first! For a brief, shining moment, everything that was wrong with the world seemed right. He never knew he could run so fast. Even his teacher and classmates seemed amazed.

Then it hit him. They weren't staring at him. They were staring at Muffin! While they were running, her ribbons had come undone and freed the biggest, most glorious afro Jaden Toussaint had ever seen in his entire life.

This day was officially the worst.

Evie patted him on the shoulder and said, "So... I'll set up that chess match for tomorrow, then?"

Oh yeah. Game on.

Chapter 5

THE ABSOLUTE WORST

"Happy" face

"Happy" Eyebrows

"Happy" Eyes

"Happy" Mouth

"Happy" Arms

"Happy" Hands

Jaden Toussaint put on a happy face on the way home, but somehow Baba still figured out that something was wrong.

Instead of saying, "What's wrong?" with plain old words, Baba used one of his superpowers and said, "You want a piggyback ride, bruh?"

Baba gave excellent piggyback rides. He zigged. He zagged. He swirled. He whirled. He went bumpity-bump-bump-bump.

When they galloped to a halt in front of their house, Jaden Toussaint had forgotten all about Muffin until she said, "Can I have a piggyback ride, too?"

PleaseSayNoPleaseSayNo
PleaseSayNo, thought
Jaden Toussaint.

Baba didn't say no. What
he said was, "What do you
think, bruh?"

   Pause.

On the inside, Jaden Toussaint was screaming,
"Nope. Non. No way, Jose!" But when it came
time to say something, what came out was very
different.

   Unpause.

"Um...I guess so..." he said.

"Thank you!" Muffin shrieked,
and she and Baba went
galloping down the block with
Sissy close behind.

Jaden Toussaint went inside
feeling angry.

Mama had put the rest of the muffins on the kitchen table for a snack. Ordinarily that would have been a good thing, but the muffins made him think of Muffin and that only made him feel angrier.

Mama saw Jaden Toussaint not eating his muffin and scooped him up onto her lap.
"You're not hungry?" Mama asked.
"I'm too angry to eat," JT said.
"Oh, I see," said Mama. "Angry is definitely one of the feelings."

"I know," JT said. He didn't mean to be rude, but sometimes it just popped out like that. He knew all about angry. He was angry with Muffin for being awesome. He was angry with Sissy for thinking Muffin was awesome. He was angry with Baba for giving Muffin a piggyback ride.

Mama continued. "I was worried you might be feeling a little jealous of all the attention Muffin has been getting. Everyone has been trying to make her feel welcome, especially since she's so far from her mother and father."

"I know," JT said. And he did know that. Mostly. I mean, he hadn't thought about Muffin being sad about being far away from her parents, but now that Mama said that part it made sense. It didn't stop him from feeling angry, though.

"Well, I keep trying to be her friend and she just keeps being..." he almost couldn't bring himself to say it, "...better than me," he finally choked out.

"Oh. I see," said Mama, holding JT a little tighter.

"What do you think we should do about that?"

"Stop her from being so awesome?" JT asked, hopefully.

"We could do that, but we wouldn't be being very good friends. Friends always want their friends to be the best they can possibly be."

That was true. Owen had never asked him to stop being awesome. Neither had Winston or Evie or Sonja.

Mama was right. He would feel awful if his friends had done that to him. He shouldn't do that to Muffin, but he still wasn't sure of what to do.
Thanks to his gigantic brain, most ideas just popped right into JT's head, but this one was a toughie. For stubborn problems like this one, there was one surefire way to kick his brain into top gear:

Turn the Page!

3 MINUTE

CRAZY
CHIPMUNK

SCOOTING X

WIGGLE WORM

WARRIOR

DANCE PARTY

Lotus

KARATE CHOP

TOOTSIE-SAN

LOOK OUT!
I THINK IT'S... YES!

Then, out of nowhere, it started. That swirly, whirly, zinging feeling he got whenever he was on the verge of a brilliant idea. And just like that, Jaden Toussaint knew what to do.

Chapter 6
THE APPRENTICE

The first thing Jaden Toussaint did was grab some books to do a little light research.

According to his research, pretty much all the coolest people in the world had learned stuff from other people. Remarkable!

So, if you can convince her to teach you, having an awesome cousin could actually make YOU more awesome. Amazing! It was such a crazy thought that JT had to try it.

BIG LIST OF
MUFFIN'S AWESOMENESS
- French
- Hellos
- Talking in front of strangers
- Traveling all over the world
- Building with Sissy
- Sports Facts
- Asking for what she wants
- Growing hair

He made a list of things Muffin was really good at. He was just about to start writing out a speech begging Muffin to be his sensei when he heard a voice at his door. It was Muffin.

"I saw you dancing earlier," she said. "It looked pretty cool. Will you teach me?"

Wait a minute. Muffin was asking HIM to teach HER something? Shocker! She was the cool one! She was supposed to be the sensei. On second thought, though, ninja dancing *was* pretty cool, and he *was* really good at it. Really, really good at it. Maybe they could both be the sensei.

"Ok," Jaden Toussaint said, "but only if you'll teach me cool stuff, too." JT and Muffin both grinned. Then they shook on it.
"Deal!" they both said.

WHAT? NO JINX?

"Let's go get a snack before we get down to work," JT suggested. "I think there are still muffins in the kitchen."

Suddenly, Muffin looked very nervous. "I've never actually eaten a muffin before," she admitted. "How do they taste?"

"What?!? You've never had a muffin?" JT exclaimed, "But that's your name!"

Muffin said something about it being hard to find muffins when they were traveling the world, but JT was too distracted to catch it all. He was trying to imagine a world without muffins. He didn't think he liked it.

"Well, do you want to try one now?" he asked. Muffin nodded.

They were all out of muffins, but Mama said they could make more. Sissy helped a little. Muffin's suggestion to swap the apples for mangoes made them even tastier, and they ate them huddled together in a fort under the dining room table. They spent the next couple of hours laughing and trading animal facts and playing toranpu cards and pretty much became lifelong friends until Mama sent them both to bed.

It wasn't until they got to JT's room that they realized they hadn't started any of the lessons. Not to worry. The night was young.

Chapter 7
THE DYNAMIC DUO

Muffin didn't master the art of ninja dancing in one night, but she did get better. And JT didn't learn all of French in one night, either, but he did learn two very important phrases:

Où se trouvent les toilettes?
(Where are the bathrooms?)
and
Avez-vous des bleuets?
(Do you have any blueberries?)

They still had the kindergarten chess match to worry about, but they also had a plan.

At first, JT had suggested that they cancel the match. After all, they were best friends now. They didn't need to try to beat each other.

Muffin pointed out that she wasn't a quitter and besides, kindergarten chess sounded like fun. She suggested that JT teach her as much as he could, and then she would try her best to play.

JT pointed out that the only real way to learn how to play kindergarten chess was to play it, and that is exactly what they decided to do.

When they entered the playground they noticed that some of the kids were wearing #TeamJT t-shirts, and some of the kids were wearing

WONDER COUSIN POWERS
ACTIVATE!

#TeamMuffin t-shirts. Evie ran over to greet
them. She was holding two of the shirts-
one that said #TeamJT and one that said
#TeamMuffin.

"Where did these shirts come from?" Jaden
Toussaint asked.

"I made them," Evie said matter-of-factly. "Mom
helped. Here's yours." Evie held the #TeamJT
shirt toward Jaden Toussaint.

"No thanks," he said. He took the one that said
#TeamMuffin instead. "I'd rather have this one."

"That's right," said Muffin, putting on the
#TeamJT shirt. "I'm on his team and he's on
mine."

And it was true. JT coached Muffin through the kindergarten chess match. He still won, but they both had fun. Muffin helped JT remember some of the hellos he forgot when he recited them for his friends. Then at snack time when JT realized he had left his snack at home, Muffin gave him half of her muffin. It was delicious.

At times like this, Jaden Toussaint couldn't help but think that with a cousin this awesome, he would always be the greatest.

Epilogue

7
~~3~~
~~2~~

Super-Sneaky ~~2~~ Fruit
Fruit Salad Muffins

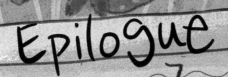

blue → (arrow)   black → (arrow)

1 - Blueberries
2 - Blackberries
3 - ~~Apples~~ Mangoes
4 - Pear
5 - Mandarin Oranges
6 - Strawberries
7 - Bananas
Flour, Sugar, butter,
baking powder, and stuff.

Step 1:
Get a grown up or certified big kid
to do the dangerous stuff.
(Note to self: Check Sissy's certification.)

Step 2:
Make sure they mix everything correctly
and don't burn themselves on the oven.

Step 3:
Enjoy

Muffin Ingredients to try:
- ~~Escargot~~
- ~~Durian~~
- Kimchi? (Muffin recommends)
- kumquats

About the Illustrator

Stephanie Parcus loves to create beautiful things. She grew up in Brazil, living off cartoons and bananas, then moved to Italy when she was 10 where she discovered Anime and Manga. Her dream to be a Pokémon trainer and her love for her dog, Fly, led her to become a veterinarian! When she's not on her farm retreat in Italy, she is traveling the world with the human who stole her heart.

THIS BOOK?

About the Author

Marti Dumas is a mama who spends most of her time doing mama things. You know - feeding ducks in parks, constructing Halloween costumes, facilitating heated negotiations, reading aloud, throwing raw vegetables on a plate and calling it dinner, and shouting, «Watch out!» whenever there are dog piles on the walk to school. Sometimes she writes, but only very occasionally and in the early morning.

You can find her at:
www.MartiDumasBooks.com

JADEN TOUSSAINT, THE GREATEST

EPISODE 1: THE QUEST FOR SCREEN TIME

Written by Marti Dumas

Illustrated by Marie Muravs

JALA

and the

WOLVES

by Marti Dumas

For crafts, recipes, and more, visit:

www.MartiDumasBooks.com

Authors love reviews.
We eat them up like
pineapples for breakfast.

Yum!

9 781943 169177